CAPSTONE PRESS
a capstone imprint

Published by Capstone Press, an imprint of Capstone
1710 Roe Crest Drive, North Mankato, Minnesota 56003
capstonepub.com

Library of Congress Cataloging-in-Publication Data
Names: Perdew, Laura, author.
Title: Your sustainable world : a kid's guide to everyday choices that help the planet! / by Laura Perdew.
Description: North Mankato, Minnesota : Capstone Press, 2025. | Audience: Ages 8 to 12 | Audience: Grades 4-6 | Summary: "It's your world, so make it a sustainable one! Learn how small, everyday choices can lead to big, positive changes for the planet-from how you use energy to what you wear, eat, and buy. Packed full of tips and relevant actions for kids to take, this is an empowering guide for any young reader looking to help the environment and make a difference. Get ready to go green and create a sustainable lifestyle, one choice at a time!"— Provided by publisher.
Identifiers: LCCN 2024003078 (print) | LCCN 2024003079 (ebook) | ISBN 9781669077190 (hardcover) | ISBN 9781669077206 (paperback) | ISBN 9781669077213 (pdf) | ISBN 9781669077220 (epub) | ISBN 9781669077237 (kindle edition)
Subjects: LCSH: Environmental responsibility—Juvenile literature. | Sustainability—Juvenile literature.
Classification: LCC GE195.7 .P48 2025 (print) | LCC GE195.7 (ebook) | DDC 640.28/6—dc23/eng/20240229
LC record available at https://lccn.loc.gov/2024003078
LC ebook record available at https://lccn.loc.gov/2024003079

Editorial Credits
Editor: Abby Huff; Designer: Heidi Thompson; Media Researcher: Jo Miller; Production Specialist: Tori Abraham

Image Credits
Capstone: Karon Dubke, 4 (left), 9 (middle); Getty Images: Nitat Termmee, 9 (right), The Good Brigade, 24, vgajic, 21; Shutterstock: AlenKadr, 25 (right), Anna Kosheleva, cover (bottom), AnnGaysorn, 4 (middle left), AS Foodstudio, 20, bokan, 17, David Tadevosian, 16 (top), Drazen Zigic, 9 (left), Elena Elisseeva, 7 (bottom left), Ernest Rose, 10, fizkes, 18, Heike Rau, 19 (left), Jazmine Thomas, 12, Julian Rovagnati, 8 (left), Lapina, 12, Lopolo, cover (top left), Magnus Binnerstam, 5, Makhh, 28-29, Maria Elisa Rol, 16 (bottom), Mehriban A, 19 (middle and right), MERCURY studio, 7 (top right), Mny-Jhee, 25 (left), Monkey Business Images, 26 (right), My Golden life, cover (top middle), myboys.me, 29 (left), Nickolay Khoroshkov, 6, Nitikan T, cover (top right), OlgaGi, 14-15, Ortis, 13, OZiBull, 25 (middle), Peyker, 7 (bottom right), Phoenixns, 26 (left), Photoongraphy, 8 (right), Rawpixel.com, 11, Red Herring, 12, Scisetti Alfio, 22, Sergey Novikov, 7 (top middle), SeventyFour, 29 (middle), showcake, 27, SmLyubov, 4 (middle right), Teri Virbickis, 7 (top left), Viktoriia Hnatiuk, 4 (right), Vlad Teodor, 14-15, wonderisland, 23, Zinkevych, 29 (right)

Design Elements
Shutterstock: Alhovik, Kwangmoozaa, Witthawas Suknantee, xnova

Printed in the United States   6286

TABLE OF CONTENTS

IT'S YOUR PLANET TOO

Every day you make a lot of choices. You choose what to wear. You pick activities to do with friends. You decide what to eat and drink. Each one of those choices has the power to help the planet.

You help when your choices are sustainable. Sustainable choices save natural resources. They protect plants, animals, and nature. These choices are good for the planet now and for the future.

Unfortunately, many activities today are unsustainable. Our activities use natural resources faster than they can be replaced. Land, water, and air are being polluted. Burning fossil fuels for energy has caused climate change. It has led to global warming. Living unsustainably harms the natural world. It harms people too.

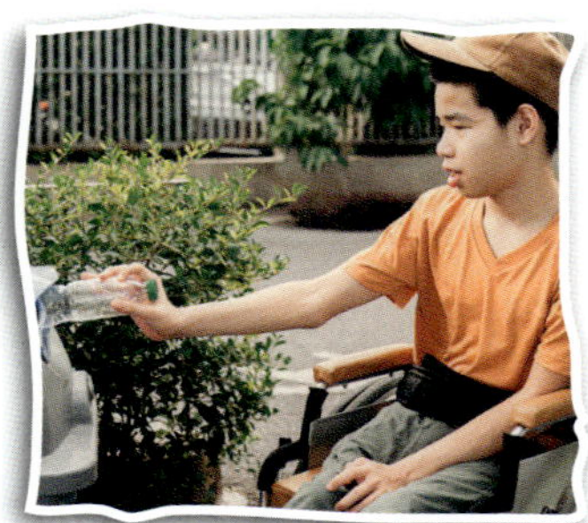

But you can help change that! By living sustainably, you play a role in protecting the planet. It starts with new choices and habits. Even small ones add up. Together they can make a big difference. Let's learn more about making sustainable choices in your everyday life!

The 5 Rs

Actions you can take every day to live sustainably!

REFUSE

Just say no! Don't buy items that create trash. Avoid wasteful packaging and single-use items.

REDUCE

Live with less. Don't buy things you don't really need. Buy new items only when necessary.

REUSE

Use items over and over again for as long as you can.

REPURPOSE

Find new ways to use an item instead of throwing it away.

RECYCLE

Put unwanted items in the recycling. That way the materials can be used again.

Energy is an important part of our lives. It keeps the lights on. It recharges our devices. It fuels our cars. But most energy is produced by burning fossil fuels such as oil, coal, and natural gas. When we use that energy, we leave a carbon footprint.

But you can reduce your own carbon footprint. How? Use less energy!

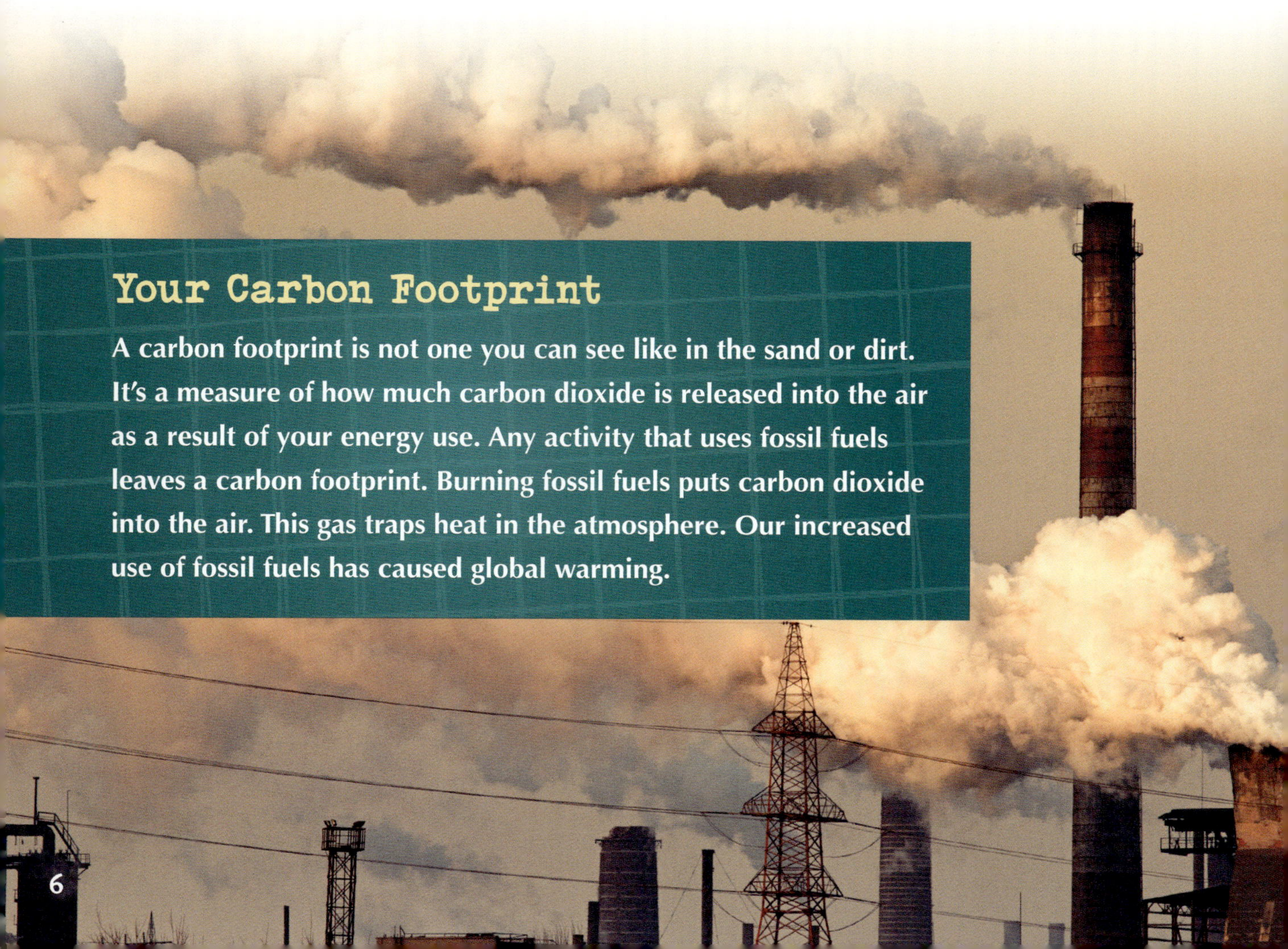

Your Carbon Footprint

A carbon footprint is not one you can see like in the sand or dirt. It's a measure of how much carbon dioxide is released into the air as a result of your energy use. Any activity that uses fossil fuels leaves a carbon footprint. Burning fossil fuels puts carbon dioxide into the air. This gas traps heat in the atmosphere. Our increased use of fossil fuels has caused global warming.

Footprints at Home

To use less, become an energy expert. Do you know what uses the most energy in a home? Heating and cooling. Save energy by fully closing the door when you go outside or come back in. You can also ask an adult about resetting the thermostat.

In the summer, keep your home warmer. To stay cool:

- Set up fans
- Make ice pops
- Wear loose, light cotton clothing
- Spray a cool water mist on your skin
- Open windows

In the winter, keep the heat lower. To stay warm:

- Wear layers
- Open curtains when it's sunny and close them after dark
- Drink hot chocolate
- Snuggle under a blanket
- Slip into thick socks or fun slippers

Power Down and Unplug

Look around your home for devices that aren't being used or that are turned off. Are they still plugged in? Believe it or not, they're using energy. Sneaky energy users include TVs and game consoles. Appliances like toasters and microwaves draw power. Make a list of what you find. Decide with your family which things you can unplug.

Chargers plugged into an outlet but not a device are also energy thieves. So unplug your device once it is at 100 percent. Then unplug the charger too.

All done in a room? Be sure to switch off the lights!

Good Green Fun

When you pick activities to enjoy with a friend, choose ones that don't require electricity. Create an art project from recycled materials. Make up a scary story. Start a neighborhood kickball game. What else can you think of?

Footprints Outside the Home

We use a lot of energy getting around town. Most people drive gas-powered vehicles, which have large carbon footprints. Taking fewer trips saves energy. You can also help by:

- Walking
- Biking
- Skateboarding
- Carpooling with friends
- Taking the school bus
- Taking a city bus, train, subway, or other public transportation

Talk with your family about what might work. These options often require more time. Think of what you can do to be ready to go. For example, want to walk to school with family or a friend? Wake up earlier. Pack your bag the night before. Or grab your own breakfast. Help out so less driving is more doable!

Being an energy expert reduces your carbon footprint. Know what else it lowers? Electric bills. And the car's gas tank won't need to be refilled as often. Using less energy helps save your family money. It helps the planet too!

MAKE A GREEN FASHION STATEMENT!

Ask for a brand-new shirt or restyle an old one? Order a fresh pair of jeans or go to a thrift store? These choices affect the planet. That's because the way most clothes are made and shipped harms the Earth.

The Truth about Fashion

Can you guess how much water is used to make a single cotton T-shirt? From growing the cotton to softening, dyeing, and spinning it, it's more than 700 gallons (2,650 liters)! Most of today's fashion is not sustainable. That's especially true with fast fashion. This clothing, including many T-shirts and jeans, is made quickly in huge amounts. It's cheap. It's meant to last for one season before being thrown away. This creates waste in landfills. And it wastes the resources used to make the clothing. Fashion leaves behind a large carbon footprint.

What's in Your Closet?

Buying fewer clothes is good for the planet. So take care of what you have. Washing clothes less helps them last longer. Put them in the laundry basket only when they are truly dirty. For example, did you wear a sweatshirt for just an hour? Stash it back in your closet. If you help do laundry, use cold water in the washing machine. It's gentler on fabric. Air drying is also gentler than using a dryer. All these actions save energy too!

Outgrown clothes still in good shape can be given to a friend or family member. Or ask about donating them. Maybe you get clothes from a sibling. Wear them proudly! You are playing a role in protecting the planet by reusing clothes and reducing the need for new ones.

Creative Clothing

Sometimes clothes just need mending. All you need is a needle, thread, and basic sewing skills. You can hide the repairs. Or show them off! Use a colorful patch to cover a hole. Stitch designs as part of the mend. Visible mending makes your clothing one-of-a-kind. And it allows you to wear each piece longer.

Another way to freshen up your style is with upcycling. This gives new life to something old. It keeps material from going into a landfill. You can:

- Turn too-short pants into shorts. Mark out the length you want with pencil. Then cut along the line.
- Add flair to an old jacket by clipping on pins or buttons.
- Use clothing you no longer wear as fabric for art projects. Be sure the clothes are clean and ask if it's okay to cut them up.

Check out books and websites with your family for even more upcycling ideas.

Get Your Green On

Upgrade a holey top or pair of pants with visible mending. Ask an adult to help you with this activity.

Supplies:

- Holey clothing
- Scissors
- Scrap fabric
- Pins (optional)
- Needle
- Thread or embroidery floss

Steps:

1. Trim any loose threads around the hole in your clothing.

2. From the scrap fabric, cut a patch that's at least 1 inch (2.54 centimeters) larger than the hole on all sides. Put the patch on the inside of the clothing and center it behind the hole. If you'd like, pin the patch down.

3. Cut a length of thread. How much you need depends on the size of the hole. But 2 feet (61 cm) is a good length to start. Push one end of the thread through the eye of the needle. Pull so it meets the other end. Holding the ends together, tie a knot.

4. Near the hole's edge, push the needle up through the back of the clothing and patch. Push it down again about 1/8 inch (0.3 cm) away. Repeat, going all around the hole. If you used pins, take them out as you sew.

5. When you've finished sewing around the hole, tie a knot on the inside of the clothing. Trim any extra thread.

6. Show off your green fashion piece! Wear it to school or when you visit a friend. Tell them about your project.

CHALLENGE

Once you have the basics, get creative! Cut the patch into a fun shape and put it on top of the hole. Make a pattern with extra stitches. Use several thread colors. Sketch how you want your next project to look so you have a plan going into the mend.

New to You

Secondhand clothes are a great choice for the planet. They are new to you but not brand-new from the store. Thrift stores sell gently used clothing. Shopping at one is like going on a treasure hunt. You never know what you'll find. Prices are often lower too. A trip to the thrift store can save your family money. Plus, it keeps clothing out of landfills!

Another option is to have a clothing swap with friends. You can lend items. Or give them away. Decide ahead of time with your family.

When it comes to fashion, less is best for sustainability. Make it a goal to have fewer items overall. Aim for them to be good quality so they last longer. To show off your green style, have a fashion show with friends!

EAT SMART

Snack time! What will you eat? Leftovers in the fridge? A bag of chips? Fresh fruits and veggies? Your food choices have the power to affect the planet!

At your next meal, think about where the food came from. It may have traveled a long way from the farm to your table. Those food miles add to your carbon footprint.

Big Farms, Big Impact

Most meat, eggs, and dairy are produced on factory farms. These farms raise lots of animals in a relatively small space. Many crops are grown on big farms too. This kind of large-scale food production feeds many people. But it harms the Earth. These places often don't use sustainable farming methods. They have huge carbon footprints. They use harmful chemicals that pollute the soil and water. Forests are cut down to grow crops and raise animals. Most products from large farms also travel long distances. The food miles from farm to processing to store to table add up!

Home Sweet Homegrown

What can you do? Encourage your family to eat local! That means buying food grown or raised nearby. Ask about taking a family field trip to a farmers market or community garden. In grocery stores, help your family look for labels that say food is from your state. Eating local doesn't just reduce food miles. Most of the time it means getting fresher food. Fresh produce is tastier. It's often more nutritious. And small, local farms often use sustainable farming methods.

Green Beans and Yams

For a greener meal, fill your plate with more fruits and veggies. Eat less meat, eggs, and dairy products. Even going meatless one day a week makes a difference. Not sure how to start?

- As a family, find books or websites with vegetarian recipes. Pick a few to try.
- Make your favorite dishes without meat. Swap beans for the meat instead. Or try plant-based meat.
- Have a family cooking challenge. Who can make the best meatless meal? Winner gets out of doing dishes for a week!

Regrow Your Food

Hang on to those vegetable scraps! They can be used to grow new food. This reduces food waste. It saves money. Plus, food grown at home is *very* local. Try regrowing the veggies below. See what you like. Then research others to try.

LETTUCE and CELERY: Cut 3 inches (7.6 cm) from the bottom of a lettuce head or celery stalk. Place the ends in a jar with 2 inches (5 cm) of water. The cut side should face up. New greens will sprout in several days.

CARROTS and BEETS: You can't regrow carrots and beets from scraps. But you can grow tasty greens! Cut 1.5 inches (3.8 cm) from the top of a carrot or beet. Place the tops cut-side down in a shallow tray of water.

GREEN ONIONS: Keep a fresh supply of green onions growing. Take a bunch of green onions. Cut an inch or so above the white part. Place the white ends in a jar with water, root-side down.

CHALLENGE

All the regrown veggies listed here (and more) can be potted in soil. After getting the hang of regrowing vegetables in jars, plant some in pots. Try starting your own small garden.

Love Your Leftovers

Stop! Don't toss that food! Instead, take steps to reduce food waste. That keeps the resources that went into producing the food from being wasted too. It keeps food out of the landfill. Food in landfills breaks down and emits gases that add to global warming. So get creative with what you have. Use over-ripe fruit in a smoothie. See if your family can use leftover beans, meat, or cooked veggies in a new dish. Repurpose those items in a salad. Add them to a pasta dish, casserole, or omelet.

Here are more ways to reduce food waste:

- Start with recommended portions and go back for seconds if you're still hungry.

- At the store with your family, avoid asking for food that isn't on the shopping list.

- Make meals together at home instead of eating out. Ask how you can help. Can you mix ingredients? Read out next steps in a recipe?

- Bring leftovers for lunch! Eat leftovers within 3-4 days. Keep track by adding the date the food was made to a bit of masking tape on the container.

- If you help unpack groceries, put new items toward the back of the fridge. Move older food to the front so it gets eaten first.

Eating sustainably is not always easy. It often requires a little more planning. And it can be more expensive. Talk to your family. Decide on one or two small changes to try. Whatever you choose, it will make a difference!

ENOUGH STUFF

Walk into any store. Shiny new products fill the shelves. They can be hard to resist. You might want the latest electronic gadget. Or a trendy toy. Or a cool bag. But each item had a long journey to get to the store shelf. And each step affected the planet.

This is where you come in. Think about the difference between needs and wants. Do you really *need* the cool bag? Or do you just *want* it? How long will you use it? Do you have an older bag you could restyle instead? Take time to think before you buy or ask for something new.

The Plastic Problem

Plastic, plastic everywhere! It's used to make straws, bottles, and bags. It's used to make toothbrushes, toys, clothing, and so much more. But it has become a huge pollution problem.

Plastic made today never goes away. That means all plastic ever made is still on the planet in some form. Over time, plastic breaks down into smaller and smaller pieces called microplastics. These tiny pieces get into the water, soil, and air. They harm wildlife and ecosystems. Scientists are still learning their effect on people.

Green Gifting

On holidays and birthdays, it's fun to receive gifts. And it's fun to give them. But gifts don't need to be bought in a store. Craft a handmade gift or repurpose an item instead. Or plan a picnic for someone. Pack their favorite foods. Or make bird feeders together using recycled materials. With an adult's help, you could plan an outing to a lake, museum, or park. Think about activities to put on your own wish list!

Use Reusables

Items that can be reused are better for the planet. Avoid single-use products. These are used only one time before they are thrown away. Many are made of plastic. Here's what you can do to help:

- **Refuse plastic straws**
- **Bring cloth bags to the store**
- **Carry a reusable water bottle**
- **Pack reusable utensils and containers in your lunch box**
- **Cut fabric scraps into reusable napkins**
- **Take a container to restaurants to pack up leftovers**
- **Make holiday or birthday decorations from cardboard boxes, old paper, or other recyclable materials**
- **Choose a cone over a cup when you go out for ice cream**
- **Skip the vending machine and pack your own snacks instead**

TIP

Use bar soap in the shower instead of liquid soap to cut down on wasteful plastic bottles.

Eco-Friendly with Electronics

People love to get the latest and greatest electronic devices. But what happens to the old ones? They often get tossed. E-waste has become a huge problem. You can help! Remember wants versus needs. Do you need a new device? Or do you just wish for one because a friend has it? Ask yourself:

- **Does your device work? If so, hold off on asking for a new one.**
- **If it's running slowly, can it be updated?**
- **If it's broken, can it be repaired?**

When you say goodbye to tech, don't throw it out. See if your family can sell or trade it. Ask about donating it. This way someone else can use the device. Or, recycle. Electronics stores often have free recycling programs.

Getting a refurbished device is another eco-friendly option. It saves your family money. It also keeps tech out of the trash for a cleaner world.

Can You Go Plastic-Free?

You are surrounded by plastic. Time to find out just how much.

Get some paper and a pencil. Start in your room. List all the items you can see that are made of plastic. Now go into the kitchen. Add to your list. Visit each room in your home. Keep taking notes.

When you're done, look at your list. Could you go one day without using any plastic? How hard do you think it would be?

CHALLENGE

Try going a week without using any single-use plastic. That includes plastic packaging.

START NOW

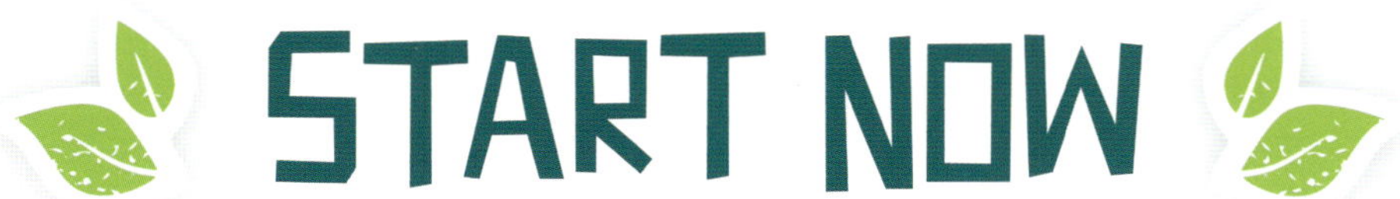

There are so many things you can do to live sustainably. It might feel like too much. Don't worry! Start with an action plan. List 3–5 things to try. Add notes next to each one about how you will make it happen. Maybe you want to eat less meat. Start meatless Mondays at home. Or maybe you want to green up your fashion. Have a clothing swap with friends. Or perhaps there is an item you were going to toss. Repurpose it instead.

Once you have a list, act! Stick to your actions for a week. Then try a second week. Soon these choices will become habits. Add more things to your action plan when you're ready.

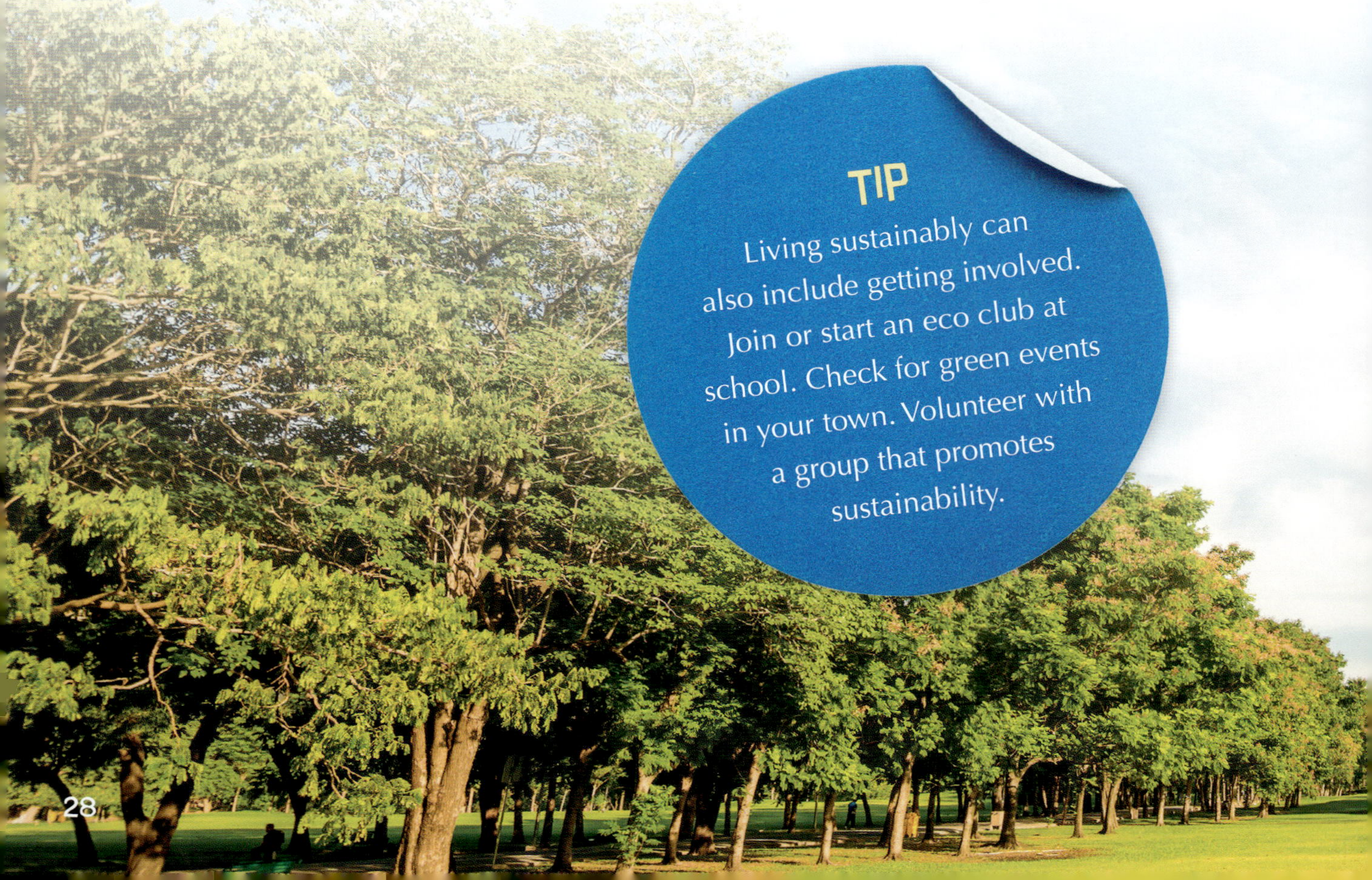

TIP

Living sustainably can also include getting involved. Join or start an eco club at school. Check for green events in your town. Volunteer with a group that promotes sustainability.

As you create new habits, keep learning about sustainability. Talk to your friends and family. Tell them about what you've learned. Explain why you're making new choices. Encourage others to join you. When a lot of people make choices that are good for the planet, it adds up to an even bigger difference.

Think of it this way. You take a reusable water bottle to school. You do this every day for the school year. That saves about 180 plastic bottles. If 30 kids in your class do the same, 5,400 bottles are saved. What would happen if everyone in your school switched to reusable bottles?

You have the power to help the planet, one choice at a time. Every sustainable choice you make promotes a strong, healthy Earth now and for the future.

GLOSSARY

carbon footprint (KAHR-buhn FOOT-print)—how much carbon dioxide is emitted as a result of one's daily activities

climate change (KLAHY-mit CHEYNJ)—the changes in Earth's temperatures and average weather patterns over a long period of time

e-waste (EE-weyst)—electronic devices that have been thrown away; this waste contains toxic materials that can poison soil and water if it ends up in a landfill

factory farm (FAK-tuh-ree FARM)—a huge farm that raises animals in large numbers

food miles (FOOD MAHYLZ)—the distance food travels from farm to factory to table

fossil fuel (FOS-uhl FYOO-uhl)—a fuel formed in the Earth from the remains of plants or animals; oil, coal, and natural gas are fossil fuels

global warming (GLOH-buhl WAR-ming)—the process of Earth's air and water heating up over time

natural resource (NACH-er-uhl REE-sors)—anything in nature that humans use, including water, trees, air, oil, and animals

refurbished (ree-FUR-bishd)—when a used product is repaired and tested in order to be sold again

sustainable (suh-STEY-nuh-buhl)—using Earth's resources in a way that keeps the planet healthy now and into the future

upcycle (UHP-sahy-kuhl)—to take an item that might usually be thrown away and turn it into something more valuable

READ MORE 

Douglas, Paul. *A Kid's Guide to Saving the Planet: It's Not Hopeless and We're Not Helpless*. Minneapolis: Beaming Books, 2022.

Minoglio, Andrea, and Laura Fanelli. *Our World Out of Balance: Understanding Climate Change and What We Can Do*. San Francisco: Blue Dot Kids Press, 2021.

Morin, Marcy, and Heidi E. Thompson. *Eco-Crafts: 40 Fun Earth-Friendly Projects*. North Mankato, MN: Capstone, 2022.

Payne, Leah. *Less Is More: Join the Low-Waste Movement*. Vancouver: Orca Book Publishers, 2023.

INTERNET SITES

The Kids Cook Monday
mondaycampaigns.org/kids-cook-monday

U.S. Energy Information Administration: Energy Kids
eia.gov/kids/

U.S. Environmental Protection Agency: Recycle City
epa.gov/recyclecity/

INDEX

ABOUT THE AUTHOR

Laura Perdew is an author, writing coach, presenter, and conservationist. She tries to make sustainable choices every day, including taking the bus, avoiding food waste, refusing single-use plastic, shopping at thrift stores for clothes, and even bringing her own reusable containers to restaurants for leftover food. And she wrote this book to help spread the word about sustainability!